The Ult

POTATO BAZOOKA

Aerosol Powered Vegetable Guns

Desert Publications

El Dorado, AR 71730 U. S. A.

The Ultimate
Potato Bazooka
Aerosol Powered
Vegetable Guns

© 1994 by Michael G. Pugliese

Published by Desert Publications
215 S. Washington Ave.
El Dorado, AR 71730
info@deltapress.com

10 9 8 7 6 5 4 3 2 1
ISBN: 978-0-87947-259-6
Printed in U. S. A.

Desert Publication is a division of
The DELTA GROUP, Ltd.
Direct all inquiries & orders to the above address.

All rights reserved. Except for use in a review, no portion of this book may be reproduced by any means known or unknown without the express written permission of the publisher.

Neither the author nor the publisher assumes any responsibility for the use or misuse of the information contained in this book. This material was compiled for educational and entertainment purposes and one should not construe that any other purpose is suggested.

Warning!!

The Publisher (Desert Publications) produces this book for informational and entertainment purposes and under no circumstances advises, encourages or approves of use of this material in any manner.

WARNING

We have strived to give you the safest *"VEGETABLE GUN"* plans possible! Please be advised however that the builder, shooter, or entity who has constructed or is in possession of any "SPUD GUN" or "VEGETABLE GUN" assumes all risk associated with the construction, use or ownership. Including any legal responsibilities, of devices detailed in this manual. *Presented for Your Enjoyment, Information, Educational, or Investment Purposes Only.*

The purchase or ownership of this manual constitutes a release of M&M ENGINEERING's responsibility for injuries property damage, or legalities incurred. Due to the wide variations in materials used in the construction, testing and use of designs detailed in this manual, it would clearly be impossible to maintain control over final construction, use or ownership.

Therefore, M&M ENGINEERING *nor the AUTHOR cannot and Will not accept any responsibility for any device detailed in this manual*

TABLE OF CONTENTS

INTRODUCTION

My experience has shown that the cheapest hair sprays almost always produced the best bang. Maybe the more expensive "stuff" had more plastics and less propellant evaporators. I know some brands listed Carbon Dioxide as a flame retardant. This takes me to the question??

Why do hair spray producers use such a highly flammable mixture of gases that they do?

It's because they boil off, that's right, they evaporate so quickly at room temperatures, almost instantly in fact, leaving the plastics dry in your hair. The Aromatic Propellants Evaporates blend with the huge amount of air around the user and they don't really present a hazard, *BUT* when confined in a SPUD GUNS chamber in the proper "air to fuel' ratio, they will produce an incredible bang.

To this day I enjoy SPUD GUN fun, you would think that in all my time spent researching and testing so many models that it would grow less impressive, exciting, interesting, it's still just plain "Good Fun". if *ANYONE* sees you shooting your SPUD GUN they will most definitely ask to try a shot. They will be so enticed by the novelty of this invention that they will most assuredly want to build one.

I'd be out shooting or testing the various models, the neighborhood would hear the first shot and come running, ready to line up for a turn, they would bring apples, pumpkin, potatos, corn, and various other garden fruits and vegetables. We created contests to see who was a crack shot, who could get the most distance, each developed their own hair spray charging and loading methods and would fight and I do mean fight any attempts by another to show how to load and shoot it. It is a heck of a lot of fun, <u>I'm *glad you* *are going to see for yourself*</u>

GOOD LUCK

MICHAEL A. PUGLIESE

This device may be restricted by laws in your area. Please check and comply with all laws.

PVC BASICS

Caution: Solvent cements and plastic pipe cleaners are highly volatile & flammable chemicals that will release airborne contaminates. Use both in a well ventilated area. Do not breath vapors. Do not allow contact with skin.

PVC pipe comes in rigid type PVC and flexible types CPVC. You will be using the white, rigid PVC (drain, waste, hot and cold water pressure systems) pipe. PVC (polyvinyl chloride) is the most common. However, CPVC (chlorinated polyvinyl chloride) is also readily available and has a similar appearance to PVC. Unfortunately, CPVC requires different cements and solvents.

NOTE: PVC plumbing pipe was not designed to be used as a gun as described in this manual. Manufacturing any Spud Gun described in this manual is inherently dangerous!

For a smooth, even cut, use a miter box and a hack saw with a fine tooth blade. It is very important to have straight even cuts.

Carefully remove rough edges, burs and surface oxidation on areas to be cemented with sandpaper or a utility knife.

Wipe your pipe and fittings with plastic pipe cleaner to remove moisture , oils, and other contaminates.

To begin cementing, apply a liberal cement coating to the outside of the pipe. Immediately apply a thinner cement coating to the inside of the fitting.

 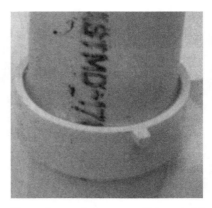

Immediately push your pipe into the fitting as far as it will go. Give it a 1/4 turn to evenly distribute the solvent cement and align the parts. Be sure to "bottom" the pipe in the fitting and keep it compressed for at least a few minutes to assure the cement sets.

If you have done everything right you will see a small bead of cement all around the joint. This indicates that you have applied the proper amount of cement.

BASIC CHEMISTRY

Propellants: Air is really made up of only <u>20% OXYGEN!</u> & 79% NITROGEN, and about 1% miscellaneous gases

Flammability Limits describe the percentages of air to gas that will support combustion, (I have computed the correct ratio for the atmospheric oxygen content of 20% by volume at 28 inches mercury and 78 degrees).
Lets look at your best fuel source, butane. At 1.55% butane and 98.45% air, butane will barely ignite, this is butane's lower ignition limit (lean). It's upper ignition limit is 8.6% butane and 91.4% air (rich). Butane is most efficient at 3% butane to 97% air. It is important for you to get the proper ratio of gas (hair spray) to air for the optimum efficiency from your "VEG GUN" and the correct ratio for butane is 31 parts air 1 part butane (31 to 1).
Propane's lower flammability limit is 2.15% and the upper limit is 9.60% with a most efficient ratio of 4% or 24 parts air to 1 part propane (24 to 1). I have included this information to show you how little hair spray is needed to get a maximum charge, and more important to show you that spraying in more hair spray will not increase your "Power", in fact a high gas to air ratio will retard or stop ignition completely. I have used the same 7 ounce can for an estimated 500 shots! The temperature at which butane will "boil off' (liquid to vapor) is +31 degrees and propane is -44 degrees, now you can see, butane will remain in a liquid state on cold winter days (remember we are igniting vapors *NOT* liquids).

Temperature: At 900 degrees butane will "self ignite", propane will self ignite at 920 degrees, your barbecue grill ignitor will produce a spark temperature of well over 3000 degrees. Chamber temperatures are below 500 degrees and last only a fraction of a second.

Heat Value is expressed in BTU's: For our purpose I have calculated the value using a one cubic foot of gas as the base value. Propane produces 2,516 BTU's, while butane produces 3,280 BTU's. *"GENERALLY"* the higher the BTU value the greater the energy release.

Ignition: Your barbecue grill ignitor is really a PIEZO-ELECTRIC-MATCH, a truly amazing little crystal. When the crystal is impacted, an electric charge is released that will jump across a half inch gap! This little wonder will produce a 3000° electric arc for years of SPUD GUN FUN.

FUEL + FRESH AIR + INGITION

=

FIRE

I don't advise the use of pure oxygen & acetylene from a welding torch. I tried it and the PVC couldn't take the pressures. In fact, it exploded my Magnum Master Blaster in my hands! I will go ahead and list a few "Saturated Hydrocarbons" you should be looking for on the ingredients listing of hair sprays and other possable fuel sources.

Butane: (WORKS WELL!) Self ignition 900 degrees, boil off 31 degrees, ratio lower 1.5% high 8.6% best ratio 3% (31 to 1)

Propane: (WORKS WELL!) Self ignition 842 degrees, boil off -43.7 degrees, ratio lower 2.1% high 9.5% best ratio 4% (24 to 1)

Ethane: Not readily available in aerosol cans as a propellant, but will burn well.

Methane: (WORKS WELL!) Self ignition 1500 degrees, boil off -260 degrees, ratio lower 4% high 14%, best ratio 10% (10 to 1)

MAPP GAS: (WORKS WELL!) Self ignition 850 degrees, boil off -25 degrees, ratio lower 3% high 11% best ratio 4% (25 to 1) *Very high temp.*

Acetylene: (WORKS WELL!) Self ignition 571 degrees, boil off -17 degrees, ratio lower 2.3% high 100% (16 to 1) *Very high temp.*

Petroleum Ether: (WORKS WELL!) Self ignition 700 degrees (Average), boil off -120, ratio lower 1% high 4% Average, (38 to 1) Average, different brands produced wide variations in testing.

Methane, Ethane, Propane, Butane, Petroleum Ether, Etc. are pure
"SATURATED UNBRANCHED HYDROCARBONS"

Other 'BRANCHED" derivatives are equally well suited for our needs. By attaching the prefix "cyclo" to designate any "SATURATED MONO-CYCLIC HYDROCARBONS" (Remember Hydrocarbons are our friends because they burn well). Another popular example is "ISO" (ISOBUTANE), which is also a combined Hydrocarbon and burns well. Pentane & Hexane, are forms of "Saturated Hydrocarbons" and will burn well especially in their pure "Unbranched" form (Propane, Butane, etc.).

WHY DOES A "VEG GUN" WORK?

In the first example on page 13 the main chamber is 3" diameter by 8" long and will have an internal area of 75.2 square inches, at an average chamber pressure experienced during testing of 20 pounds per square inch. This area will produce a total chamber pressure of 1508 pounds *(Do not confuse total pressure with pounds per square inch).* 1508 sounds like a very powerful force but it is the TOTAL FORCE acting on the *entire internal area of the combustion chamber* useful only to engineers interested in design strength requirements. These pressures will act upon the 1.5 inch diameter potato, forcing the potato out of the 1.5" PVC tube (BARREL) with a force of 46.7 pounds per square inch (1.5" X 46.7 = 70.05).

In a split second 70.05 pounds force will act on the area of a 4 ounce potato and that's what makes this device work so well.

Chamber length and diameter play an important part in flame front speed and the resulting chamber pressures. As a general rule, keep your chamber short. Increase your chamber diameter to gain power.

CAN WE INCREASE CHAMBER SIZE TO INCREASE POWER?

I have included a chart on page 13 showing the "predicted" increase in chamber pressures. But, the fact is, it would be unwise to build one as large as the ones in the chart because of the way the chamber pressures rise so quickly when the chamber diameter and length are increased. A device of this size could only be built from heavy wall steel tubing, not plastic, and would resemble a mortar more than a harmless and fun "VEG GUN" so I won't describe it's construction at all. Over 4" X 16" chambers are definitely not recommended and are in fact dangerous.

Chamber Size			Total Pressure			Pounds Per Square Inch	
DIAMETER	*LENGTH*	*SQUARE INCHES*	*AVERAGE CHAMBER PRESSURE*	*TOTAL*	*PRESSURE*		*ACTING PRESSURES*
3" x 8" = 75.2			x 20LB. = 1508			1.5 x 12 = 1002	
3" x 10" = 94.4			x 20LB. = 1880			1.5 x 12 = 1253	
3" x 12" = 112.8			x 20LB. = 2256			1.5 x 12 = 1504	
4" x 10" = 125			x 20LB. = 2500			1.5 x 12 = 1666	
4" x 12" = 150			X 20 LB. = 3000			1.5 x 12 = 2000	
4" x 14" = 176			x 20LB. = 3506			1.5 x 12 = 2333	
6" x 10" = 188			x 20LB. = 3760			1.5 x 12 = 2506	
6" x 12" = 222.5			x 20LB. = 4512			1.5 x 12 = 3008	
8" x 12" = 301.2			x 20LB. = 6024			1.5 x 12 = 4016	
8" x 14" 351.4			x 20LB. = 7028			1.5 x 12 = 4685	

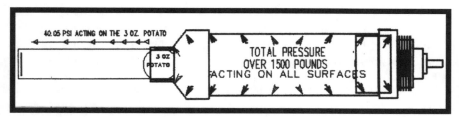

The internal chamber area must contain the blast long enough for your potato to be expelled from the barrel releasing the pressure. This PVC piping is actually over 300% stronger than design strength requirements need to be. Or that's what my computer aided design program tells me and we know the real world is always just a little different than what the computer tells us.

It is important to make sure that the largest percentages of Volatile Saturated Hydrocarbons or their derivatives are present in your spray. Looking at the ingredients listing you will see the chemicals listed in the order of amounts present in the spray. Highest percentages listed are first and lowest percentages listed last.

Take for example "ABC" Hair Spray, whose list contains 16 separate chemical compounds with SD Alcohol 40 and Butane listed as the first 2 ingredients. At first inspection it seems that this is a perfect choice for a "VEG GUN" propellant, it has a total of 2 propellent liquids or gases, but it is really the worst choice! The remaining 14 chemical ingredients although present in lesser proportions, will retard your flame front (in fact many producers are putting in flame retardants) also of importance is that you will be exposed to hazardous gasses after the burning of these remaining 14 chemicals, which include toxic plastics and fragrances that don't smell so nice after burning.

Now lets look at "XYZ" hair spray ingredients listing, SD Alcohol 40, Isobutane, Propane, Ethyl Ester of PVM\MA Copolymer, Butane, Dimethyl Stearamine, Dioctylsebacate, Fragrance. Now you have 8 ingredients, but look at them closely, SD Alcohol, Isobutane, Propane are the first and largest percentages of Aromatic propellants listed, plus in smaller quantities are Ethyl Esters of PVM\MA Copolymer and Butane, both very Aromatic propellants although one is a plastic compound held in suspension. "XYZ" hair spray has a total of 5 proponent gases and 3 remaining ingredients. Look for Saturated Hydrocarbons in the greatest percentages and remaining chemicals (PLASTICS) in the least percentages.

Automotive starting fluid and brake cleaner in a spray can is a powerful propellent. It's only draw back being it requires a strict control of your air to fuel ratios. They are now putting Carbon Dioxide, to retard combustion, in many brands so you should look at the ingredient listing very carefully.

WARNING: I tried using a oxygen acetylene cutting torch set to mix the gases and insert them into the combustion chamber. It launched the potato completely out of sight!! I felt the PVC plastic tubing expand in my hands! ! ! The pressure gauge, Piezo-Match and end cap blew completely out, in fact the end cap shattered, spewing extremely high temperature fire on me and everywhere!!! The resulting noise left my ears ringing for days! *NEVER USE PURE OXYGEN OR AN OXYGEN RICH MIXTURE! This design is intended to use 20% atmospheric oxygen*

I tested all the above fuel sources with good results. Hair spray was the most consistent and easiest to use of all gases tested. Don't use pure oxygen as shown in the small pocket sized welding torch. The use of propane, butane and mapp gases provide exceptional results. I used a common propane torch to charge my chamber, it went like this. Since the gas and air are mixed in the torch head a perfect ratio is already developed for you. Simply insert the tip of the torch just inside the chamber turn on the gas for about 5 seconds and try it.

ECONO BLASTER

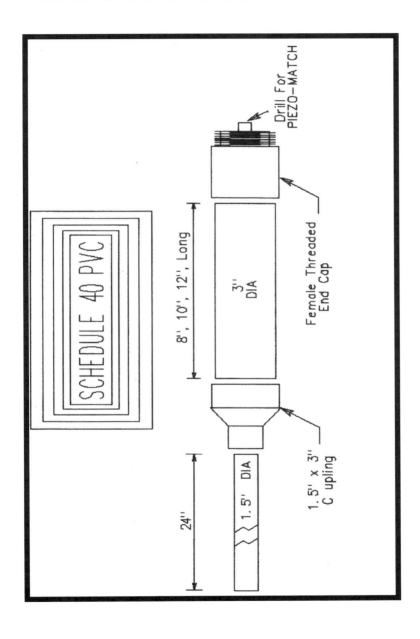

LIST OF MATERIALS

SCHEDULE 40 PVC OR BETTER

3" DIAMETER PVC TUBING 8", 10" OR 12" LONG (see chart on page 13)

3" FEMALE END CAP WITH THREADED "CLEAN OUT" PLUG

3" TO 1 1/2" REDUCING COUPLING

1 1/2" DIAMETER PVC TUBING 24" long

1 1/2" DIAMETER TUBING 3" LONG (sharpen one end to form a potato cutter)

1 REPLACEMENT BARBECUE GRILL IGNITOR

1 SMALL CAN OF PVC CEMENT

1 SMALL CAN OF PVC SOLVENT (PVC cleaner/primer)

1 PIECE OF SAND PAPER OR STEEL WOOL PAD

1 CAN OF HAIR SPRAY

1 TAMPING ROD (28" long broom handle)

YOUR CHOICE OF VEGETABLES (potatoes, apples, beets, tomatoes, etc.)

TOTAL COST FOR EVERYTHING SHOULD BE AROUND $ 20.00

This has got to be the lowest cost SPUD GUN possible. I tried every conceivable PVC combination and this is without a doubt the best design for simplicity in assembly, and it's also a very safe sized design. It typically launches potatoes and apples over 100 yards without any problem. Only a very little shot of hair spray is needed (about a 2 seconds blast) or better still give it a one second blast and immediately follow with another one second blast. This seems to swirl the hair spray and air in the chamber to mix better, go ahead and be creative in loading and charging your SPUD GUN. Just don't do anything crazy. I have used potatoes, apples, corn, beets, pumpkin, squash, and more for projectiles.

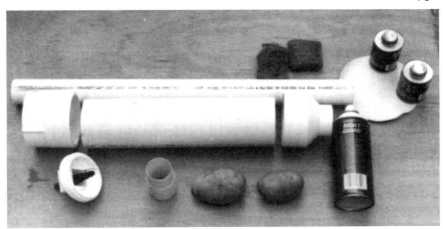

Showing the entire assembly needed to complete one Econo-Blaster and fire your first shot. Start by cutting the tubing into the proper lengths. Grind or file away your bullet cutter, sand the burs and surface oxidation away from the areas to be cemented. The front of the barrel should receive a sand papers "rounded muzzle". This will ease the loading of your vegetables greatly. I also drilled the hole and installed the PIEZO-MATCH in my end cap.

The author fires a shot from the best and least expensive design. I really liked this model the most!

Drill a hole snugly enough to stop gas leakage into your threaded end cap as shown. Install your PIEZO-MATCH and screw into place. If you cannot find a threaded PIEZO-MATCH you can use a clip held model by cutting the retaining clips off and then epoxy the PIEZO-MATCH into place.

Begin cementing the 3"
female threaded end cap
onto the 3" chamber tub-
ing.

Cement the 3" to 1 1/2" reducer coupling onto the
chamber tube.

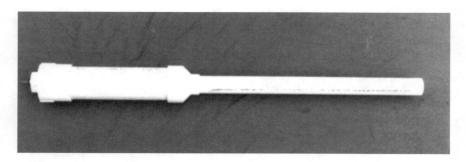

Cement the 1 1/2" X 24" to 32" barrel tube onto the reducer coupling
and you're finished. Now let your spud gun dry overnight to assure the
cement is dry. Do Not Use The Spud Gun no matter how eager you
are. The cement must be completely cured. After final assembly, while
the cement cures, keep the spud gun in a warm area standing as straight
up as possible and do not disturb until the next day!

MASTER BLASTER

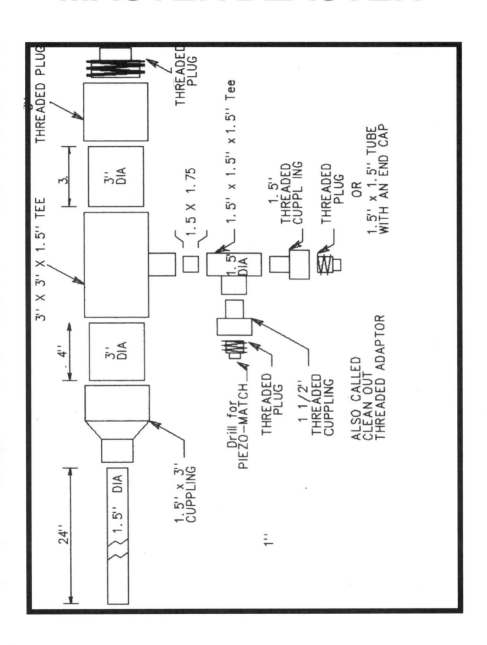

LIST OF MATERIALS

SCHEDULE 40 PVC OR BETTER

3" DIAMETER PVC TUBING 3" LONG

3" DIAMETER PVC TUBING 4" LONG

3" X 3" X 1 1/2" PVC TEE

3" END CAP WITH THREADED "CLEAN OUT" PLUG

3" to 1 1/2" REDUCING COUPLING

1 1/2" DIAMETER PVC TUBING 30" LONG

1 1/2" DIAMETER TUBING 3" LONG (sharpen one end to form a potato cutter)

1 1/2" X 1 1/2" X 1 1/2" PVC TEE

1 1/2" END CAP

1 1/2" PVC TUBING 1 3/4" LONG

1 1/2" THREADED END CAP

1 REPLACEMENT BARBECUE GRILL IGNITOR

36" BROOM HANDLE (or 1/2" plastic piping to be used as a ram rod for loading)
YOUR CHOICE OF VEGETABLES potatoes, apples, beets, tomatoes, etc.)

This design will "launch" a 1 1/2" X 3" potato a maximum distance of 200 yards! I love to launch my potatoes straight up, that way I can shoot them over and over again. Oh man, do they go up! An additional 2" of barrel (32" total) has shown some improvements in range over the standard 30" model detailed above, I have even used 34" barrels on longer chamber designs with good success.

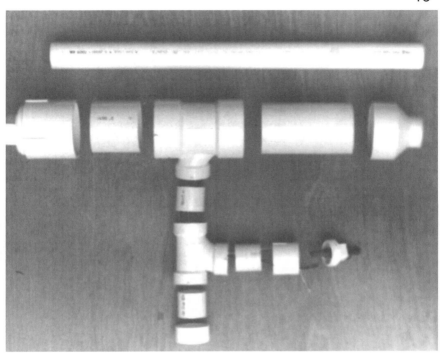

Showing the entire assembly needed to complete the Master-Blaster. Start by cutting the tubing into proper lengths. Grind or file away your bullet cutter, sand the burs and surface oxidation away from the areas to be cemented. The front of the barrel should receive a sand papers "rounded muzzle". This will ease the loading of your vegetables greatly. I also drilled the hole and installed the PIEZO-MATCH in my end cap.

You can use a spark plug type of PIEZO-MATCH as shown in the Econo-Blaster or this variation shown. The proper gap should be 3/8 inch or less. You can glue into place or use a nut and bolt or sheet metal screw to hold it in place.

Drill a hole snugly enough to stop gas leakage into your threaded end cap as shown. Install you PIEZO-MATCH and screw into place. If you cannot find a threaded PIEZO-MATCH you can use a clip held model by cutting the retaining clips off and then epoxy the PIEZO-MATCH into place.

Begin by gluing the 3" female threaded end cap onto the 3" PVC chamber tubing. You can cut the chamber tubing from 3" to 6" long.

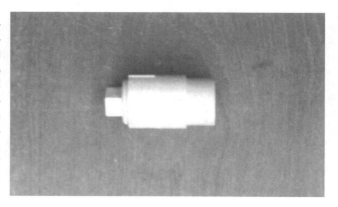

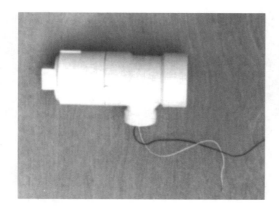

Cement the large tee onto the chamber tube. The joints should be touching as shown in the photo for maximum strength. This butting of the joints will increase strength by effectively doubling the tubing thickness.

Cement the 3" X 4" long chamber pipe into your large tee. Make sure that the previously cemented joints are not loosening.

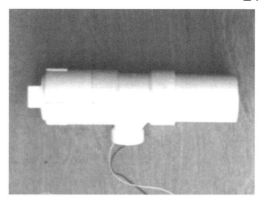

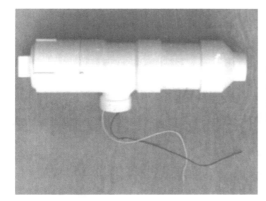

Cement the 3" X 1 1/2" reducer coupling on the chamber tubing. Look over the entire assembly again.

Cement the 1 1/2" X 1 3/4" handle tubes into the larger tee. Make sure that the PIEZO-MATCH wires are feed through the fittings as we progress in construction from this point on.

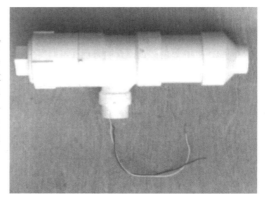

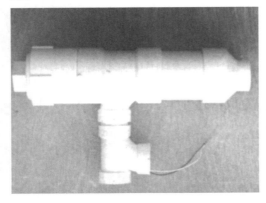

Cement the small tee onto the chamber tube's tee. You sould postion the fittings exactly as shown.

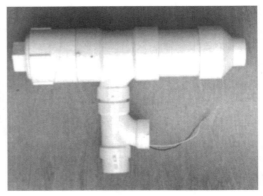

Cement the 1 1/2" X 1 3/4" tube onto the bottom of the smaller tee.

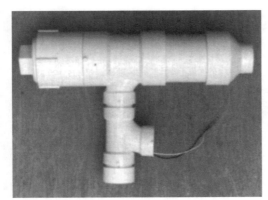

Cement the 1 1/2" end cap onto the bottom of the smaller tee.

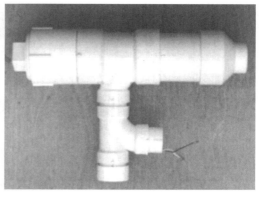

Cement the 1 1/2" X 1 3/4" trigger tube onto the smaller tee.

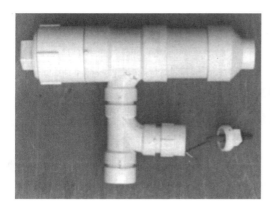

Cement the 1 1/2" female threaded end cap onto the smaller tee. Connect the electrode wires to the PIEZO-MATCH and screw the threaded plug into the handle.

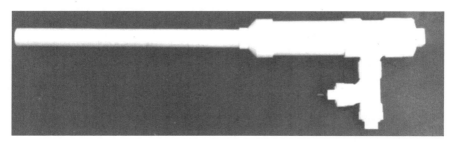

Cement the 1 1/2" X 24" barrel tube onto the reducer coupling and you're finished. After final assembly keep the SPUD GUN standing as straight up as possible and do not disturb for one entire day. DO NOT use the SPUD GUN no matter how eager you are. The cement must be completely cured.

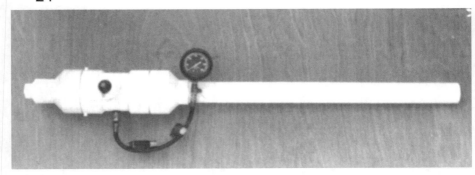

The pressure test gun above is designed very differently from the designs you have seen and the reason is the cone shaped rear end cap produced a greater flame front speed and the threaded end plug is smaller and stronger. The threaded end plug and the PIEZO-MATCH are the weak areas in spud guns.

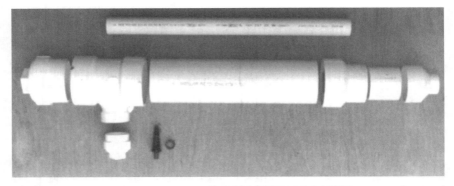

An experimental MEGA MAGNUM MASTER BLASTER measuring a super big 4" X 24" chamber and a 32" long barrel. Although it failed to provide any significant power improvements, when it's bulky size was taken into consideration.

At this point in my research I discovered flame front speed proved to be essential to consistent superior performance. Increase the chamber diameter instead of lengthening your chamber for more power!

MAGNUM MASTER BLASTER

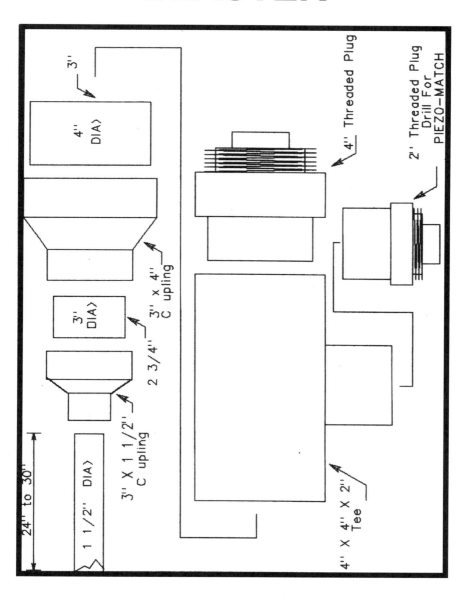

4'' DIA>

3''

4'' Threaded Plug

2'' Threaded Plug
Drill For
PIEZO-MATCH

3'' x 4''
Coupling

3'' DIA>

2 3/4''

3'' X 1 1/2''
Coupling

1 1/2'' DIA>

24'' to 30''

4'' X 4'' X 2''
Tee

LIST OF MATERIALS

SCHEDULE 40 PVC OR BETTER

4" DIAMETER PVC TUBING 3 1/2" LONG, 14" MAX.

4" x 4" x 2" PVC TEE

4" MALE THREADED END CAP with THREADED CLEAN OUT PLUG

4" TO 3" REDUCING COUPLING

3" TO 1 1/2" REDUCING COUPLING

1 1/2" PVC TUBING 24" LONG

1 1/2" PVC TUBING 4" LONG (potato cutter)

2" MALE THREADED END CAP with THREADED CLEAN OUT PLUG

1 REPLACEMENT BARBECUE GRILL IGNITOR KIT

YOUR CHOICE OF VEGETABLES note; potatoes, apples, and other *hard vegetables* are the best choice, simply because most other vegetables are torn to pieces by the violent acceleration in the barrel and subsequent impact onto the outside atmosphere. Caution: This design really puts a potato out there, so be very sure of your field of fire. This model if launched straight up will BLAST the potato so high it will smash into pieced when it hits the ground, it will break slate roofing and dent cars, be careful!
A hair spray shot of about 3 to 5 seconds is a good starting point. Because of the larger 4" chamber you will be able to purge the chamber of burnt, spent gases by forming a loose fist and inserting it into the chamber once or twice, this will push out the spent gases and suck in a charge of fresh air. This design will produce a substantial muzzle flame and the resulting sound signature will be equally violent, I definitely recommend ear, eye and proper protective clothing for all SPUD GUNS but this model mandates their use!

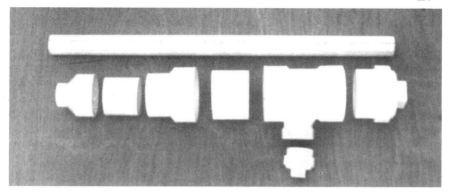

Showing the entire assembly needed for the assembly of one MAG-NUM MASTER BLASTER. Start by cutting the tubing into proper lengths. Grind or file away your bullet cutter, sand the burs and surface oxidation away from the areas to be cemented. The front of the barrel should receive a sand papers "rounded muzzle". This will ease the loading of your vegetables greatly. I also drilled the hole and installed the PIEZO-MATCH in my end cap.

Be-fore-warned! The chamber pressure on this size of a PVC SPUD GUN is getting very high. I had several mishaps during testing. I recommend you approach with great caution and with personal safety foremost in mind. In one incident I was totally engulfed in very high temperature flames when an end cap gave way during testing.

Begin by drilling the smaller 2" threaded end cap to fit the PIEZO-MATCH and install with a threaded nut or epoxy to secure in place.

Cement the 4" male threaded end cap into the 4" X 4" X 2" tee. Keep the 2" entrance angled away from the end cap.

Cement the 2" threaded male end cap into the tee. Protect your PIEZO-MATCH in a safe place until the PVC cement is completely dry.

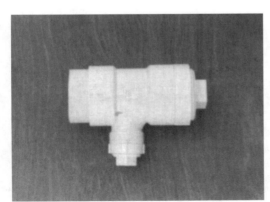

Cement the 4" X 3 1/2" PVC chamber tube into the tee.

Cement the 4" to 3" reducer onto the 4" chamber tubing.

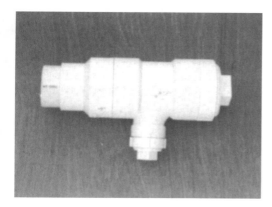

Cement the 3" X 3 1/2" chamber tubing into the reducer.

Cement the smaller 3" X 1 1/2" reducer onto the 3" chamber tubing

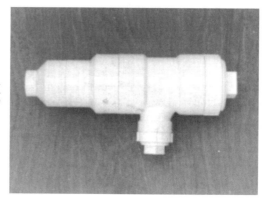

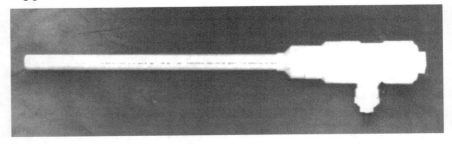

Cement the 1 1/2" X 24" to 34" barrel into the smaller 3" to 1 1/2" reducer coupling and you're finished. After final assembly keep the SPUD GUN standing as straight up as possible and do not disturb for one entire day. DO NOT use the SPUD GUN no matter how eager you are. The cement must be completely cured.

Here is the author firing a shot from the MAGNUM MASTER BLASTER

(and it really is!). Please be very cautions as this is really the upper limits on the PVC plastic SPUD GUN. I had the most fun with this model because of it's incredible POWER. In fact, it demanded the use of ear protection and other safety equipments like welding gloves. You see I had a few mishaps on the larger models and was a little weary of them. If you only use hair spray and one potato you should be OK? I experienced mishaps when I used other propellants which were not really as good or reliable as hair spray. In the beginning I pictured a wide variety of other possible gases you could use, but good old hair spray is the overall top performer.

SPUD BLASTER PISTOL

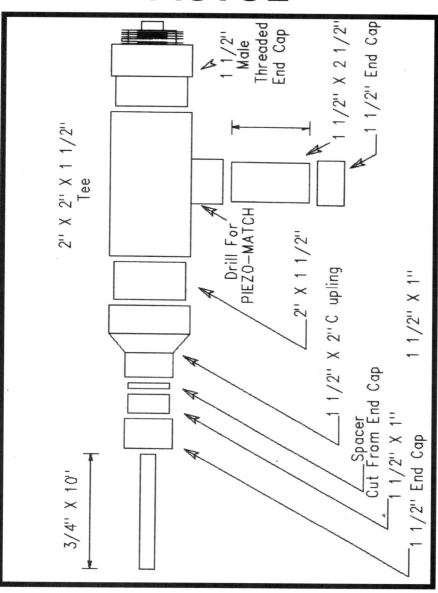

1 1/2" Male Threaded End Cap

1 1/2" X 2 1/2" End Cap

1 1/2" X 2 1/2"

2" X 2" X 1 1/2" Tee

Drill For PIEZO-MATCH

2" X 1 1/2"

2" X 2" Coupling

1 1/2" X 2" End Cap

Spacer Cut From End Cap

1 1/2" X 1"

1 1/2" X 1"

1 1/2" End Cap

3/4" X 10'

LIST OF MATERIALS

SCHEDULE 40 PVC OR BETTER

1 1/2" DIAMETER PVC TUBING 1 1/2" LONG (NOT NEEDED FOR SILENCER VERSION)

1 1/2" DIAMETER PVC 2 3/4" LONG (Handle)

1 1/2" DIAMETER 10" LONG (Silencer version, barrel shroud)

2" x 2" x 1 1/2" PVC TEE

2" DIAMETER PVC TUBING 1 1/2" LONG

2" TO 1 1/2" REDUCING COUPLING

2" MALE THREADED END CAP

3/4" DIAMETER PVC TUBING 11" LONG

3/4" DIAMETER PVC TUBING 2" LONG (Bullet cutter)

3 1 1/2" END CAPS

1 REPLACEMENT BARBECUE GRILL IGNITOR KIT

You will be shooting a 3/4" potato about 50 yards. A longer barrel was the winning ticket but it wasn't really a pistol any more, I'll let you experiment. The chamber is so small that very small amounts of gas are needed.

To charge your chamber:
(1) Remove the threaded end cap and flush some fresh air into the chamber.

(2) Now hold the pistol in one hand, the hair spray in the other, point the spray just out side the chamber opening about 10" away and begin spraying. Pass the mist over the chambers open end smoothly in one continuous motion.

(3) Quickly screw the end cap on, your chamber is sealed and ready.

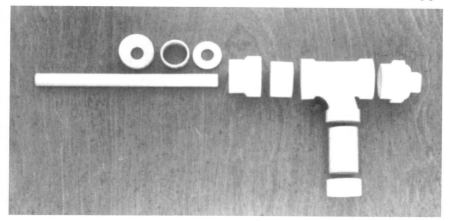

Showing the material needed for the assembly of one SPUD PISTOL. Start by cutting the tubing into proper lengths. Grind or file away your bullet cutter, sand the burs and surface oxidation away form the areas to be cemented. The front of the barrel should receive a sand papers "rounded muzzle". This will ease the loading of your vegetables greatly. I also drilled the hole and installed the PIEZO-MATCH in my tee.

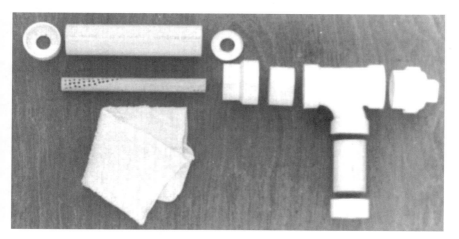

Showing the materials needed for the assembly of one SILENCED SPUD PISTOL. The truth is this pistol is not that loud that it would require a silencer. This is just fun information to read, look at and study, right? Does this give you any ideas about modifying your other SPUD GUN's

Cut a washer from an end cap. Now drill a hole for your barrel in your second end cap.

Cement the threaded end cap onto your tee. Make sure the trigger hole is facing forward.

Cement the handle onto the tee.

Cement the end cap onto the handle tube.

Cement the chamber tube onto the tee.

Cement the reducer coupling onto the chamber tube.

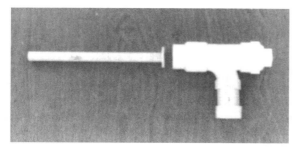

Cement the barrel onto the washer you made from an end cap. Keep everything as straight as you possibly can from this point on.

Cement the washer onto the reducing coupling and cement the spacer tube onto the reducer coupling.

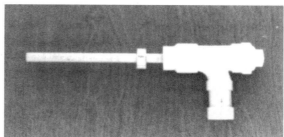

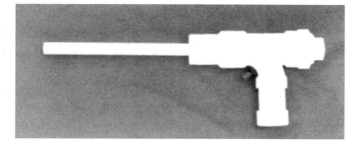

Cement the end cap that has been drilled for the barrel onto the spacer. Center everything up for the last time. Dry overnight.

The completed SPUD BLASTER. I had a lot of fun getting the 3/4" potato out of my yard.

SILENCED SPUD BLASTER

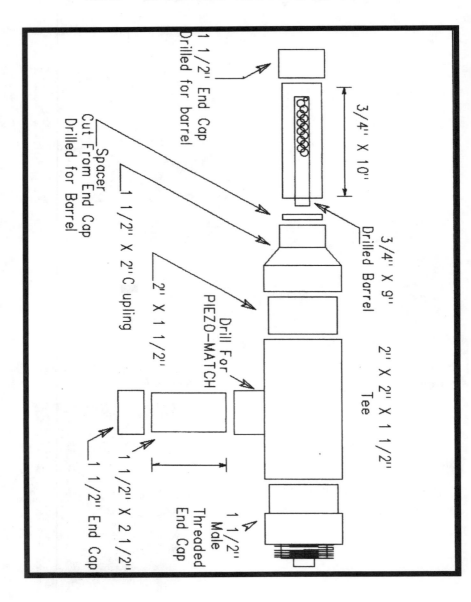

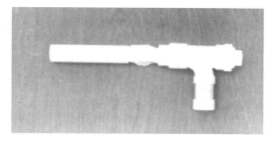

Instead of using a 1 1/2" X 1 1/2" spacer use a 10" long piece of 1 1/2" tube. By wrapping a wash cloth around your drilled barrel you will very effectively silence the muzzle report.

The finished SILENCED SPUD PISTOL. It is a real pleasure to shoot and is quite accurate for it's size.

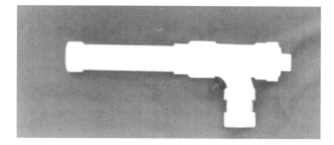

The author testing the SPUD PISTOL. I really enjoyed writing and researching this manual for you. I definitely had fun testing and experimenting with different versions. GOOD LUCK!

Look forward to a manual I plan to print in the near future detailing the construction of a "MAD MAX" type of survival weapon that I have invented using similar principles. This 'one' has never been thought of, or at least it's not in print, YET!

I knew, when I printed this manual, it would be a big success. If I only had insight like that with the stock market I would be rich, rich, rich. I have sold 4 reprints in the 2 years since it's first printing with each reprint doubling in quantities printed. I have had hundreds possibly thousands of calls from builders with questions on use and construction. I finally began to write down their questions, and the answers I gave, so I could rewrite and improve my manual with each reprinting. You are now reading my first 'New and Improved' reprint. I didn't delete information or pictures, but instead added to them.

Almost 90% of the calls were from people who didn't read the manual but instead built a SPUD GUN right from the pictures. They bought the most expensive hair spray which was packed full of carbon dioxide to retard flame. They didn't allow their SPUD GUN cement to dry over night, so the hair spray softened the cement destroying the cements strength. They raced blindly along and wondered why their SPUD GUN "won't fire". PLEASE READ THE MANUAL.

OTHER IMPORTANT QUESTIONS
?If I increase chamber size should I increase barrel length?
My answer is YES in most cases. Allow for the extra burn time required with larger chambers by increasing barrel length to a maximum of 36".
?I have spark (using a 3/8" gap), I used hair spray and it still won't light and launch a potato?
Clean out the chamber with a dry cloth, swing the SPUD GUN with it's end cap off to get fresh air into the chamber. Spray a 1 second blast directly into the chamber and quickly screw on the end cap. Point in a safe direction and depress the trigger. If it doesn't fire spray in 1 more second of hair spray, try it again. If it still doesn't fire, try another brand of hair spray. Most people use way to much hair spray thinking it will work better, it won't. READ THE MANUAL.
?I live outside of America and I don't know where to get PVC?
First let me thank you for your letters, I see the postage you paid on the envelopes and I read every one thoroughly.
Most Plastic water piping is sold with a pressure rating of 200 psi and that will withstand any hair spray ingition. Many letters describe a gray plastic piping used in South America that they say works very well. I can't say either way, but my overseas readers say it's cheap and works fine.

GOOD LUCK!
MICHAEL A. PUGLIESE

Notes

Other Books Available From Desert Publications

PRICES SUBJECT TO CHANGE WITHOUT NOTICE

Desert Publications

800-852-4445

215 S. Washington Ave Dept. - BK - 259

Shipping & Handling 1 item $6.95 - 2 or more $9.95

El Dorado, AR 71730 USA